Jacob Chimuca

# Biological Treatment of Wastewater

Jacob Chimuca

# Biological Treatment of Wastewater

**Dynamic Membrane Anaerobic Bioreactor
applied to wastewater treatment**

ScienciaScripts

**Imprint**
Any brand names and product names mentioned in this book are subject to trademark, brand or patent protection and are trademarks or registered trademarks of their respective holders. The use of brand names, product names, common names, trade names, product descriptions etc. even without a particular marking in this work is in no way to be construed to mean that such names may be regarded as unrestricted in respect of trademark and brand protection legislation and could thus be used by anyone.

Cover image: www.ingimage.com

This book is a translation from the original published under ISBN 978-620-6-76022-1.

Publisher:
Sciencia Scripts
is a trademark of
Dodo Books Indian Ocean Ltd. and OmniScriptum S.R.L publishing group

120 High Road, East Finchley, London, N2 9ED, United Kingdom
Str. Armeneasca 28/1, office 1, Chisinau MD-2012, Republic of Moldova, Europe
Printed at: see last page
**ISBN: 978-620-7-66934-9**

# *DYNAMIC MEMBRANE AEROBIC BIOREACTOR APPLIED TO WASTEWATER TREATMENT*

*JACOB FORTUNA JOSÉ CHIMUCA CATARINA SIMONE*
*ANDRADE DO CANTO*
*JOSÉ TAVARES DE SOUSA VALDERI DUARTE LEITE*
*WILTON SILVA LOPES*

**SUMMARY**

Membrane bioreactors have been widely used in the biological treatment of wastewater. The membranes used in this type of technology are produced from synthetic, organic or inorganic materials. However, membranes can also be formed from the deposition of solid particles, colloids, polymeric materials, as well as microbial cells and flocs, on an inert support during the filtration process and, when coupled to a bioreactor, they become a single system called a Dynamic Membrane Bioreactor (DMBR). This type of bioreactor, while retaining the solids and microorganisms present in the system, removes organic material that is easy and difficult to degrade, which reduces treatment costs and makes it advantageous over conventional membrane bioreactors (BRM). In Brazil, this technology is relatively new and still little explored. Therefore, this review aims to evaluate the performance of BRMD's in anaerobic wastewater treatment systems. In addition to the advantages and disadvantages presented by this type of system compared to conventional BRMs (micro and ultrafiltration), the phenomenon of fouling and its implications and theories that explain the formation of the dynamic layer are described. Finally, some of the challenges that this technology still needs to overcome in order to consolidate itself as a safe and robust tool for the biological treatment of domestic and industrial wastewater are pointed out.

**Keywords:** dynamic membrane, fouling, wastewater treatment challenges, bioreactor performance.

# CONTENTS

# 1. INTRODUCTION

Membranes are barriers used in both separation and biocatalytic processes, in which they usually act as supports for enzymes, cells or microorganisms. Since separation and biocatalysis processes are closely related to biochemical processes, membranes can be coupled to bioreactors, becoming so-called Membrane Bioreactors (MBRs). BRMs perform the important function of separating the biocatalyst from the reagents and products and thus carry out the separation and biocatalysis processes in a single step (DAVIS, 2010).

Membranes are generally produced from synthetic materials of an organic or inorganic nature. Organic membranes are based on cellulose and modified organic polymers such as cellulose acetate, polypropylene, polyethylene, polycarbonate, among others (LADEWIG and AL-SHAELI, 2017). The raw materials for inorganic membranes are ceramics and metals. Ceramic membranes are normally used in industrial applications; however, when it comes to wastewater treatment, they are economically prohibitive due to their high commercial value. Metal membranes, on the other hand, have very specific applications, which can also be related to biotechnological treatment processes (ZHANG et al., 2005).

However, membranes can also be formed by depositing colloids or suspended solid particles, such as microbial cells and flocs, on a porous material, which acts as a support for the formation of a biologically active layer during the filtration process (MARCINKOWSKY et al., 1966; YAMAGIWA et al., 1994; PILLAY et al,

1994; WU et al., 2005). In this case, there is the so-called Dynamic Membrane (DM) which, like synthetic membranes, can also be coupled to bioreactors and used in separation and biotransformation processes. The use of Dynamic Membrane Bioreactors (DMBRs) in ultrafiltration processes began in the mid-1980s (ERSAHIN et al., 2012). A decade later, research began into the (aerobic/anaerobic) treatment of wastewater (YAMAGIWA et al., 1994; PILLAY et al., 1994). Over the last 30 years, the results of numerous studies have shown removal efficiencies for suspended solids, biochemical oxygen demand (BOD), chemical oxygen demand (COD), total organic carbon (TOC), nitrogen and phosphorus comparable to those presented by reactors operating with conventional ultra and microfiltration membranes, thus confirming the feasibility of using dynamic membranes in the biological treatment of wastewater (FAN and HUANG, 2002; KISO et al., 2005; AN et al., 2009; MA et al., 2013a,b; ERSAHIN et al., 2016 a,b; AYOL et al., 2021).

Interest in dynamic anaerobic membrane bioreactors (DAMBs) has been growing since 2007 and, although these systems have shown promising results in terms of treatment, research has generally aimed to optimise the operating conditions of the systems (AN et al., 2009; ZHANG et al., 2010; ERSAHIN et al., 2012). However, despite the good results already found, much still needs to be clarified about this type of technology, which has shown itself to be a viable alternative to established systems.

This literature review aims to evaluate the performance of BRAnMD's in wastewater treatment systems, describe the advantages and disadvantages of applying this type of system, the implications of fouling and the theories that explain the formation of the dynamic layer during the filtration process. Finally, some of the challenges that need to be overcome in order to improve the functionality of this technology, making it more robust and reliable, are described.

# 2. DYNAMIC MEMBRANE BIOREACTORS

The dynamic membrane, also known as a secondary or biological membrane, is usually formed on an inert, inexpensive material with a pore size ranging from 10 to 200 µm and with almost negligible intrinsic resistance (ALIBARDI et al., 2014; GUAN et al., 2018ab, WANG et al., 2020). The support material can be polyester, polypropylene, polyethylene terephthalate (PET), stainless steel, cotton, nylon, non-woven fabric (TNT), activated carbon sponge and even a synthetic membrane, generally used when the solution to be filtered contains particles of suspended solids such as microbial cells, flocs and colloids (LI et al., 2018b; POLLICE and VERGINE, 2020; AYOL et al., 2021).

Because the pore sizes of the supports are relatively large, these materials do not act as filtration media, but in fact as supports. Organic and colloidal particles, which normally block the pores of micro- and ultrafiltration membranes, are retained in the filtration layer of the MD, thus preventing the formation of fouling on the support material (Figure 1). For this reason, modules with cheap, macroporous materials are used for the MD filtration process, instead of relatively expensive membranes with extremely small pores (ALIBARDI et al., 2014; ERSAHIN et al., 2017).

Figure 1 - Schematic showing membrane filtration under transverse flow: (a) MF or UF membranes; and (b) MD formed on macroporous support material.

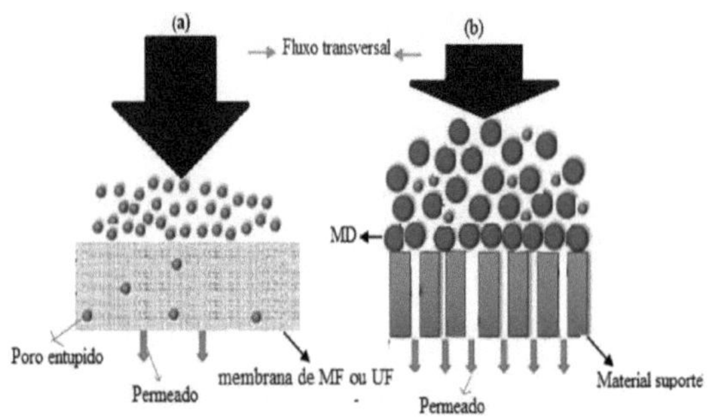


Source: Adapted from Pajooh (2018).

Compared to MF and UF membrane bioreactors, BRMD's have advantages such as: (i) 1 to 4 times higher flux; (ii) lower investment costs, since the support material used to deposit the dynamic layer is relatively cheap; (iii) high anti-pollution potential, which allows fouling to be easily removed and flux restored;
(iv) low filtration pressure, allowing the effluent to flow by gravity, without the need for pumps and therefore saving energy, as well as (v) the ability to retain helminth eggs (WU et al., 2005; ERSHAIN et al., 2012, 2017; ALIBARDI et al., 2014; LIU et al., 2016; CHIMUCA et al., 2020; POLLICE and VERGINE, 2020; VERGINE et al., 2021). DM is also a This is a promising technology for removing low-density, non-biodegradable microparticles such as microplastics, which are highly present contaminants in wastewater. These microparticles are not easily removed by conventional sedimentation

and therefore result in higher operating and maintenance costs for the unit processes following secondary treatment (LI et al. 2018). Despite their advantages, including economic ones, dynamic membrane bioreactors are unstable during their practical application. In addition, parameters such as cell retention time (CRT), hydraulic detention time (HDT), extracellular polymeric substances (EPS), permeate flow and the morphological properties of the sludge (sedimentation, dehydration, flocculability, hydrophobicity), which are a function of the feed, temperature and dissolved oxygen concentration in the medium, if not strictly controlled, can negatively affect the bioprocess (VERGINE et al., 2021). However, one of the main disadvantages observed in membrane systems is the phenomenon known as fouling, defined as the unwanted deposition of particles, colloids, macromolecules and salts on the surface of the membrane (external fouling) and/or inside its pores (internal fouling), which causes a reduction in the flow through the membrane and, consequently, a loss in process performance (MENG et al. 2009; FANE et al., 2011; LADEWIG and AL-SHAELI, 2017). Despite being an extremely complex phenomenon, since it involves a number of factors, its control can be achieved by modifying the membrane surface or optimising the operating conditions depending on the bioprocess in progress, for example (FANE and FELL, 1987; WEI et al., 2006).

It can be said, however, that the continuous replacement of fouling-causing fine particles on the surface of micro and ultrafiltration membranes is responsible for more than 80 per cent of the filtration resistance resulting from the formation of irreversible fouling. On the other hand, the same particles, as well as the PMS (proteins and polysaccharides) and inorganic substances present in the medium, help the active biomass to form the dynamic layer, which has the dual function of filtering and biologically treating wastewater, thus becoming an important and promising tool in the field of sanitation. In addition, using this technology it is also possible to obtain an effluent with a quality similar to that of bioreactors using MF and UF membranes, while significantly reducing the costs of the process (TOWNSEND et al., 1989; NOOR et al., 2002; LIU et al., 2009; MAHAT et al., 2018; YANG et al., 2020). The dynamic membrane seems more suited to anaerobic processes, due to the slow growth of microorganisms and the milder hydrodynamic conditions resulting from the absence of aeration. The configuration, which can be internal or external, plays an important role in the effectiveness of the process. However, studies have shown that the internal configuration appears to be more attractive, particularly in terms of the rapid formation/regeneration of DM, which directly affects the quality and efficiency of the treatment (ZHANG et al., 2010; ERSAHIN et al., 2017; QUEK et al., 2017; SUN et al., 2018).

It is also worth mentioning that the application of a dynamic membrane can increase the substrate retention capacity in the dynamic membrane anaerobic bioreactor (BRAnMD) and subsequently stimulate an increase in the microorganisms responsible for the hydrolysis step in the digestion process due to an increase in the amount of hydrolytic enzymes (such as protease and $\beta$-glucosidase), which can result in changes in the organic matter degradation pathways. In addition, bacteria that are essential for the degradation of refractory organic matter, and which normally exhibit a low growth rate, can be retained and accumulated in the dynamic layer, allowing for greater degradation of refractory substances into more easily biodegradable by-products, such as volatile fatty acids (LIU et al., 2016). Most studies have been carried out on a bench and pilot scale and, although these systems have shown promising results on a par with systems using synthetic (commercial) membrane bioreactors, research has generally been aimed at optimising the operating conditions of the bioreactors. Therefore, studies aimed, for example, at eliminating pathogens and recovering products still need to be better investigated (AN et al., 2009; ZHANG et al., 2010; ERSAHIN et al., 2012; ALIBARDI et al., 2016; CHIMUCA et al., 2020).

## 2.1. Dynamic membrane formation

Several theories have been formulated to explain how solids and other particles contained in the liquid medium agglutinate and deposit on a support material. Among them is the Extended Derjaguin - Landau - Verwey - Overbeek Theory (XDLVO Theory), which aims to explain the mechanisms of adhesion of biocatalysts on the surface of a support (AZEREDO et al., 1999; VILINSKA and RAO, 2011; RUAN et al., 2020). This theory can also be applied to the process of forming a dynamic membrane, which physically occurs according to the processes of adhesion and cohesion.

In the initial stage of MD formation, the sludge flows from the bioreactor onto the clean surface of the support. This process is dominated by the interactions between the support and the sludge flocs, and is defined as the adhesion behaviour of the sludge (adhesion process). In the next stage, called the DM maturation stage, most of the membrane surface is covered by sludge flakes. At this stage, the interactions between the sludge flakes approaching the support and the flakes already deposited replace the interactions that predominate in the initial stage and begin to control the formation of the dynamic layer. In this way, the interactions between the sludge flakes, i.e. flake-flake interactions, are defined as the cohesive behaviour of the sludge (cohesion process) (YU et al., 2019b).

The formation of DM can also be described as a dynamic process developed under hydrodynamic and thermodynamic forces. Hydrodynamic forces include permeate drag, inertial, shear and gravitational forces. Thermodynamic forces, on the other hand, are represented by the physical-chemical interactions that take place over a short distance

(d) between the surface of the support and the sludge flakes, and are defined as the Lifshitz - van der Waals (LW), Lewis acid - base (AB) and electrostatic double layer (EL) forces (HONG et al., 2013). Normally, the drag force of the permeate, which is induced by the flow through the membrane, can transport the sludge flakes to the surface of the support. The total forces (XDLVO forces), which are the sum of the LW, AB and EL components, are attraction forces that may be responsible for the attachment of the sludge flakes to the surface of the support. Figure 2 shows the analysis of the microcosmic force of a sludge flake located very close to the surface of the support. In general, the XDLVO theory makes it possible to predict, in a practical way, how the process of forming a dynamic membrane takes place, while also making it possible to optimise the properties of the sludge and the operation of the system.

**Figure 2:** Analysis of the microcosmic force of a sludge flake located very close to the surface of the support.

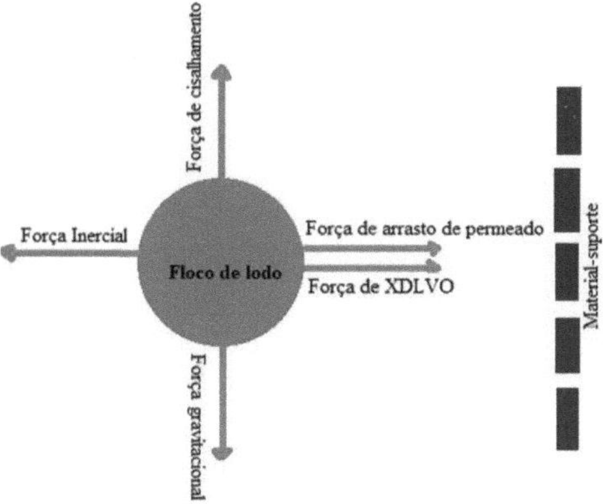

Adapted from YU et al. (2019b).

In addition, considering the fundamentals of fluid mechanics, the drag force of the permeate plays a very important role in the dynamic membrane formation process when the support material has relatively large pores and hydrophilic characteristics. In this case, when the initial permeate flow is high, larger sludge flakes with low relative hydrophobicity are formed. rapidly deposited on the surface of the support, causing the resistance to filtration to increase, despite the dynamic layer having a low specific resistance. However, as the membrane forms, the pores in the support and the flow through the membrane decrease, while the adhesion forces increase due to the accumulation of smaller sludge flakes and the production of SPEs. It is at this stage that the dynamic layer becomes more compact and has a higher specific resistance. Thus, the movement of the sludge flakes deposited on the support material includes the drag force of the permeate and the shear force on the surface of the support material, which respectively control the convection velocity and the reverse transport velocity, both of which depend on the permeate flow. In this way, the deposition rate of the sludge flakes is governed by the net velocity in the direction of the support material, since there is a balance between the convection velocity and the reverse transport velocity, although the role of the shear force is insignificant in the process (SUN et al., 2018).

After the complete formation of the MD, a good retention capacity of the small flocs is achieved and the permeate flow is substantially reduced, associated with a decrease in the porosity of the MD and, consequently, the formation of a more compact membrane. These factors contribute to a higher retention capacity and an increase in transmembrane pressure (TMP) and specific resistance during the filtration process.

Before the formation of a stable DM, the support material cannot induce solid-liquid separation, because the DM consists mainly of a cake layer formed from the bioreactor sludge and a gel layer formed from the sludge metabolites, mainly extracellular polymeric substances (SPE), which develop during filtration. In addition, the quality of the effluent from MD bioreactors in the first few hours, or even days, is extremely low and the loss of sludge is significant. However, the quality gradually increases with prolonged operation. In other words, sufficient time must be allowed for the filtration process to take place. occurs properly, so that a stable DM can form for effective solid-liquid separation, as well as sludge retention by the system (SIDDIQUI et al., 2019).

However, the macroscopic indication of DM formation is a marked reduction in permeate turbidity, along with a drastic decrease in flux through the membrane (HU et al., 2016a; YU et al., 2019b). The maturation stage of DM is also characterised by a continuous decrease in permeate turbidity, a reduction in permeate flux and an increase in filtration resistance. According to Hu et al. (2020), the time required for the formation of a stable DM layer is in the range of 10 to 25 days, during long-term operation (approximately 30-50 days) and under a flow rate of 1.0 to 8.0 $L.m^{-2}.h^{-1}$. Thus, there are many theories based on forces, speeds and the properties of the microbial flocs and the support material, as well as the characteristics of the operation and the final effluent. However, there seems to be a consensus that DM is formed in two stages: the initial stage and the maturation stage.

## 2.2. Fouling in Dynamic Membrane Bioreactors

Fouling can be understood as a set of phenomena capable of causing a drop in membrane performance over time. These phenomena arise when working with a solution or mixed liquor, and their consequences can be totally or partially irreversible. Fouling causes restriction and occlusion of the membrane pores through the deposition of organic and inorganic particles, which results in an increase in filtration resistance and, consequently, a reduction in permeate flow (MARTINEZ-SOSA et al., 2012). In addition, the phenomenon has distinct characteristics and natures, i.e. it can be reversible and removable or not and be organic, inorganic or biological in nature, as shown in Table 1. In the case of dynamic membrane bioreactors used to treat wastewater, fouling can be affected by factors such as: characteristics of the support material (type of material and pore size), biomass characteristics (solids concentration, extracellular polymeric substances, soluble microbial products, floc structure and size) and operating conditions (transmembrane pressure, permeate flow, hydraulic detention time, cell retention time, cleaning frequency) (LIAO et al., 2004).

**Table 1:** Types of fouling and their characteristics.

| Type of fouling | Description | Reference |
|---|---|---|
| Removable | Arising from particles that bind weakly to the membrane and are associated with the formation of the dynamic layer. | LADEWIG; AL-SHAELI, 2017 |
| Non-removable | Usually caused by heavily adhered scale during the filtration process. | LADEWIG; AL-SHAELI, 2017 |
| Organic | Caused by the deposition of proteins, polysaccharides, humic acids and other organic substances from the sludge, such as SPE and PMS, on the membrane or inside its pores. | LIN et al., 2013 |
| Inorganic | Caused by the deposition, on the membrane or inside its pores, of inorganic compounds such as struvite (MgNH4PO4.6H2O), ammonium phosphate and potassium (K2NH4PO4) and calcium carbonate (CaCO3). | KANG et al., 2002 |
| Biofouling | Formed due to à deposition e growth of microorganisms on the surface of the membrane. | MEABE et al.,2013 |
| Reversible | This occurs when deposition occurs only on the surface of the membrane and is therefore easy to remove through purely physical processes. | PARK et al., 2015 |
| Irreversible | This occurs when deposition is due to the adsorption of organic and inorganic substances in the pores of the membrane. | PARK et al., 2015 |
| Colloidal | This refers to the fouling of membranes with suspended and colloidal particles such as microorganisms, polysaccharides, sludge, lipoproteins, lignin, sulphur and sulphides, among others. | LADEWIG; AL-SHAELI, 2017 |

The type of material the support is made of strongly affects dynamic membrane bioreactors, although there is no clear information on the real influence of this parameter on the performance of the dynamic layer itself. However, it has been observed that supports made from cotton promote greater efficiency in the removal of organic matter and suspended solids, as well as favouring the formation of a thicker and more resistant dynamic layer. On the other hand, nylon, as well as being low cost, allows for the complete removal of organic matter and suspended solids. fouling and the restoration of the structure through simple physical cleaning, such as backwashing (LI et al., 2018b; POLLICE and VERGINE, 2020; YANG et al., 2020; AYOL et al., 2021).

In addition, fouling is more severe on hydrophobic supports than on hydrophilic ones, although changes in hydrophobicity often occur due to other characteristics of the support, such as pore size and morphology, which make the correlation between hydrophobicity and fouling more difficult to assess (YU et al., 2005). Since biomass has hydrophobic characteristics at different levels, its affinity and, consequently, its adherence to the surface of the support material with hydrophobic characteristics is more intense than that observed with the use of hydrophilic materials (CHANG and JUDD (2002).

In conventional membranes, if the particle size is larger than the pore size, filtration is blocked or restricted due to fouling. In dynamic membrane systems, the pore size of the support material is also a key parameter for filtration performance through the dynamic layer, even when the operating conditions favour the rapid formation of a more stable dynamic membrane (HE et al., 2005; POLLICE and VERGINE, 2020).Results indicate good filtration performance when the dynamic membrane is formed on a support material with a pore size varying between 10 and 100 μm (KISO et al., 2000; FAN and HUANG, 2002; CHU et al., 2008).

For materials with a pore size of less than 5 μm, 99% of the pores can be blocked (CAI et al., 2018). However, if the support material has pores in the order of 10-40 μm, the formation of the dynamic membrane is relatively fast and the effluent generated is of high quality. However, the pores in the support are easily blocked, decreasing the flow through the membrane and therefore increasing the need for more frequent backwashing. On the other hand, when the pore size of the support material is greater than 100 μm, there is a larger filtration area, higher fluxes and lower resistance to filtration, but the effluent generated is of high quality. The formation of the dynamic membrane requires a longer time. Also, with larger pore sizes, it may not be easy to obtain high quality effluent due to greater membrane instability (sludge loss). Therefore, in order to determine the most appropriate support material, pore size is critical in terms of dynamic layer formation time, flux and pollutant removal (YURTSEVER et al., 2020; AYOL et al., 2021).The total solids present in a biological reactor can be divided into: dissolved solids (smaller than 0.001 μm), colloidal solids (0.001 to 1.0 μm) and suspended solids (larger than 1.0 μm), and they affect the behaviour of the membrane differently. Colloidal and soluble particles, for example, contribute to the total or partial blockage of pores and can affect the membrane's performance irreversibly. In general, with the other operating parameters kept constant, an increase in the concentration of total solids leads to a decrease in permeate flow due to an increase in the thickness of the dynamic layer. However, the thickness of the layer can be minimised by improving the hydrodynamic conditions of the system and thereby reducing the deleterious effects of fouling (DERELI et al., 2012).

Extracellular polymeric substances (EPS) consist of a complex mixture of proteins, carbohydrates, polysaccharides, DNA, lipids and humic substances that make up the matrix of flocs and biofilms. On the other hand, soluble microbial products (SMP) are a mixture of organic compounds resulting from metabolism and cell death during the substrate mineralisation process. These products are easily absorbed by the dynamic layer and can block the pores and reduce the permeability of the support material, as well as influencing the structure and porosity of the dynamic layer formed (NG and HERMANOWICZ, 2005; LIANG et al., 2007; PRETEL et al., 2013).

Total pore blockage by SPE and PMS is the main fouling factor in micro- and ultrafiltration membrane bioreactors, which normally does not occur in BRMD's due to the fact that the pores of the support are relatively small.and therefore more difficult to block completely (LIU et al., 2012; MENG et al. 2017). However, in BRMD's, the blockage of the pores of the support material can be caused by larger microbial flocs, consisting mainly of bacteria coated with SPE and PMS and which cause the microbial flocs to form a dense structure with few narrow channels for the passage of permeate (GUAN et al., 2018a,b).

Biological flocs, in turn, consist of heterogeneous structures made up mainly of microorganisms (bacteria, protozoa, fungi, among others), organic and inorganic particles, extracellular polymeric substances and soluble microbial products. It is worth noting that the function of SPE and PMS in floc formation is to increase the viscosity of the liquid, favouring extracellular enzymatic activity and, consequently, cell aggregation (LEE et al., 2003). The structure of the floc is strongly related to the operating conditions in dynamic membrane bioreactors. Thus, if the tangential velocity conditions are too high, they can reduce the size of the floc and lead to the formation of more compact cake layers on the surface of the support material. It is therefore important to optimise the tangential flow velocity over the support material so that it favours the deposition of the cake on the support, but does not cause a high increase in the concentration of solutes and colloids of a size comparable to or smaller than the pore size of the support material. This is a strategy capable of minimising the effects of fouling and the deterioration of permeate quality (WANG et al., 2020).

Sludge with stable structures forms and protects large flocs from being broken or disaggregated by external forces (e.g. shear force) or turbulence. In addition, sludge flocs with a stable structure potentially reduce the release of SPEs from the biological flocs, which relieves blockage and promotes the formation of reversible fouling in the DM. In other words, the good aggregation capacity of the sludge inn dynamic membrane has lower fouling potential and higher permeability through the dynamic layer formed (YU et al., 2019a).

Other parameters that have a strong influence on fouling in dynamic membrane bioreactors treating wastewater are permeate flux and transmembrane pressure. The permeate flux, defined as the volume that permeates through the membrane per unit time and per unit area, is influenced by the quality of the affluent, the stability of the operation and the properties (such as the type of material and pore size) and cleaning frequency of the support material. It is also influenced by operating conditions, such as the transmembrane pressure (pressure gradient that serves as the driving force for transport through the membrane) and the mixing regime in the bioreactor. It is worth mentioning that low transmembrane pressure values allow for much higher fluxes in dynamic membrane bioreactors, compared to those observed in synthetic membrane bioreactors (METCALF & EDDY, 2007; POLLICE and VERGINE, 2020).

When comparing filtration between dynamic membranes and synthetic membranes, it can be seen that higher values of transmembrane pressure are only achieved when the porosity of the dynamic membrane is reduced by the excessive accumulation of solids or by other phenomena related to sludge growth, such as the overproduction of PMS or SPE, which usually results in the occurrence of fouling (HU et al., 2016a,b; POLLICE and VERGINE, 2020).

Another important concept is critical flow, which consists of the highest permeate flow value below which fouling does not occur or is negligible (HAN et al., 2005). On the other hand, above the critical flux, fouling occurs due to a greater tendency for particles to be dragged along during filtration, causing them to settle more quickly and not allowing the hydrodynamic conditions to prevent the membrane pores from blocking. It is important to emphasise that the critical flow can be determined in two different ways. One is by keeping the flow constant and observing the behaviour of the transmembrane pressure. Thus, the critical flow will be established when the PTM begins to increase. Another way is to keep the transmembrane pressure constant and measure the permeate flow. In this way, the critical flow will be defined when the permeate flow begins to decrease (HAN et al., 2005).From a practical point of view, for the treatment of wastewater in dynamic membrane bioreactors, a high cell retention time (CRT) associated with a low hydraulic detention time (HDT) promotes a higher concentration of sludge and enables efficient treatment in a short space of time. However, such conditions can generate more intense fouling problems, which implies that the TDH and TRC variables may be indirectly linked to the impacts caused by fouling in this type of system (TRUSSELL et al., 2006). However, the time needed to form a stable dynamic layer can take up to 50 days, depending on the permeate flow used (ERSAHIN et al., 2014; ALIBARDI et al., 2014). For this reason, cell retention and hydraulic detention times must be rigorously defined so that a stable and efficient DM is formed in a short space of time.

Studies have shown that very low hydraulic detention times (1 h, for example) lead to an increase in transmembrane pressure due to an increase in the concentration of soluble microbial product, which leads to poor process performance and poor effluent quality. In addition, very low TDHs allow for a greater accumulation of inorganic substances in the dynamic membrane, which results in severe fouling caused by the shorter reaction time (YANG et al., 2020).

Reducing the cell retention time, on the other hand, can cause an increase in filtration resistance and, consequently, PTM. Huang et al. (2019) found that for a TRC of 5 days, a layer of biofilm similar to a thin, compact gel (porosity = 27.5%), with relatively high filtration resistance (approximately $4.9 \times 10^{11}$ m$^{-1}$ ), while a thick, porous biofilm layer (porosity > 60%), with lower filtration resistance (<$2.5 \times 10^{9}$ m$^{-1}$ ), was formed at TRCs greater than or equal to 20 days. In addition, it was observed that higher TRCs reduce the production of SPEs, but increase the population of protozoa in the biofilm, which increases the porosity of the biofilm and reduces biofouling in the support material. Although the increase in filtration resistance and the reduction in permeability in dynamic membrane bioreactors are much lower compared to those observed in synthetic membrane bioreactors (approximately two orders of magnitude lower), fouling control becomes even more necessary when there is a severe decline in flow or a rapid increase in PTM. And in order to at least minimise such deleterious effects during the wastewater treatment bioprocess, one must first try to define suitable operating and hydrodynamic conditions or even carry out pre-treatment of the suspension that will feed the membrane system. However, of all the factors that control fouling, cleaning techniques are the only ones that have the function of recovering the permeate flow and can be carried out by chemical, physical or biological methods, whenever necessary (POLLICE and VERGINE, 2020).

Chemical cleaning is carried out using strong acids, caustic products and/or oxidising agents that restore the membrane's performance almost completely. However, the cleaning solution must have a pH compatible with the pH range supported by the support material (YU et al., 2003; CUI et al., 2003; BÉRUBÉ et al., 2006).

Schneider & Tsutiya (2001) emphasise the importance of the formulation of the cleaning solution, as fouling is rarely caused by a single type of material deposited on the membrane support. To remove inorganic deposits, acid cleaning is generally used, while for the to remove organic deposits and biofilms, the formulation of the solution must be alkaline. It is therefore recommended to use different formulations alternately, as repeated applications of the same solution can result in the selection of a resistant biofilm. In general, manufacturers of support material indicate commercial formulations that are compatible with their products. In the event of the possibility of secondary contamination caused by chemical products, physical cleaning is used, such as backwashing, a method widely used in membrane systems. Backwashing pushes particles adhered to the pore structure into the liquid and partially removes the cake formed on the surface of the support mesh. The frequency and flow rate of backwashing are related to the operating conditions and the characteristics of the effluent to be treated. The method normally uses air, water or the permeate itself, which is pumped in the opposite direction to permeation (OGNIER et al., 2004).It is worth mentioning that the frequency of backwashing does not significantly affect the performance of MD bioreactors and therefore an automated cleaning protocol for the support material can be easily applied at full scale for MD bioreactors (WANG et al., 2020; AYOL et al., 2021). Cleaning can also be carried out by: (1) using nitric oxide (in low concentrations) to induce biofilm dispersion, (2) enzymatic hydrolysis of the extracellular polymers formed and (3) breaking the bond between the biofilm and the membrane surface, which can be promoted by the addition of bacteriophages. Although these techniques are still being developed for use on a bench scale, there is a possibility that they could also be applied on a real scale (PARK et al., 2015).Fouling can also be controlled depending on the configuration of the module in which the membrane is located. In this sense, in externally configured systems, a high tangential flow velocity is maintained to limit the accumulation of organic and inorganic substances responsible for fouling. On the other hand, in the submerged configuration, fouling is controlled by spraying with gas or backwashing with tap water or the effluent itself (HUANG et al., 2011; SMITH et al., 2012).

However, mathematical modelling can define the mechanism of fouling formation during the filtration process through the dynamic membrane. Ruth, B.F., in the mid-1930s, established the Theory of Filtration through the cake layer, providing the basis for the definition of Modern Filtration Theory (IRITANI and KATAGIRI, 2016).

Although in MD filtration processes the fouling particles play a decisive role in the development of the dynamic layer, few studies attempt to understand specifically how these phenomena occur through model-based analyses. Furthermore, the use of mathematical modelling to elucidate the mechanism of DM formation, together with the effect of changes in operating conditions (PTM and suspended solids concentration) on the model response and its interpretation, is still limited.The filtration blockage laws describe both membrane pore blockage and dynamic layer formation, as well as allowing the increase in filtration resistance to also be evaluated, since this is directly responsible for the decline in flow over time and under constant pressure conditions, or the increase in pressure over time under constant flow conditions. The fact is that fouling can be described by more than one model that explains the different membrane pore fouling mechanisms (pore constriction, pore fouling and the formation of a cake layer on the membrane surface), which can occur successively or simultaneously. In this regard, Iritani and Katagiri (2016) defined the four models shown in Table 1 and Figure 1.

**Table 1.** Mathematical models for evaluating the evolution of fouling over time.

| Filtration Model | Model equation | Locking constant |
|---|---|---|
| Complete lockdown | P 1= | Kb (1.s ) -1 |
| | P0        1- Kb .t | K (1.m−3 ) s |
| Standard lock | P $\left.\right]$  K . J .t |  |
| | $\left.\right]^{-2}$ =$\left[$ 1-S 0  P0   2 | Ki (1.m ) -3 |
| Intermediate blocking | P = exp(K . J .t ) iP        0 |  |
| | 0 | Kc (s.m ) -6 |
| Filtration through the layer pie | P = 1+ K . J .t P     c     0 0 |  |

Where: P is the PTM (Pa); P0 is the initial PTM (Pa); J0 is the flow (L h$^{-1}$ ); Ki is the complete blockage constant; KS is the standard blockage constant; Ki is the intermediate blockage constant; Kc is the filtration constant through the cake layer.

In the complete blockage model it is assumed that each particle that reaches the surface of the support material obstructs the passage of a pore, i.e. completely closes a pore on the surface of the support material. Furthermore, a particle never settles on another particle that has been previously deposited, i.e. there is no overlapping of particles. The permeate flow through the unblocked pores is not affected, so the reduction in permeate flow is proportional to the reduction in the surface area of the support material corresponding to the unblocked pores. It is worth emphasising that this type of fouling occurs when the particle size is larger than the pore size of the support material. Consequently, pore blockage only occurs on the surface of the support material and not inside its pores (Figure 3a). The standard blockage model assumes that blockage occurs inside the pores of the support material and thus gradually reduces the pore volume until it is completely blocked. The reduction in pore volume is proportional to the reduction in permeate volume (Figure 3b). It is also considered that the pore has a constant diameter and length throughout the thickness of the support material. Furthermore, as this type of obstruction is caused by particles smaller than the pores of the support material, fouling becomes independent of the pore volume. transverse flow velocity. Thus, the steady-state flow is equal to zero and the counter-diffusion of solutes from the surface of the support material to the centre of the solution does not occur.

The intermediate blocking model considers that the pore of the support material is not necessarily blocked by a single particle, i.e. several particles have the same probability of depositing on top of each other, causing the pore to block. Once again, blockage inside the pores is not taken into account. Furthermore, this type of fouling occurs when the particles are similar in size to the pore size of the support material (Figure 3c).Finally, in the cake layer filtration model, the particles are larger than the pore size of the support material and the particle concentration is high. Therefore, there is no pore blockage. In this case, particles are deposited on the surface of the support material, forming a first layer. Subsequently, a new layer is formed on top of the first and so on (Figure 3d), characterising the adhesion process that occurs during the formation of the dynamic layer.

**Figure 3** Schematic representation of the fouling mechanism: (a) complete blockage; (b) standard blockage; (c) intermediate blockage and (d) blockage through the cake layer.

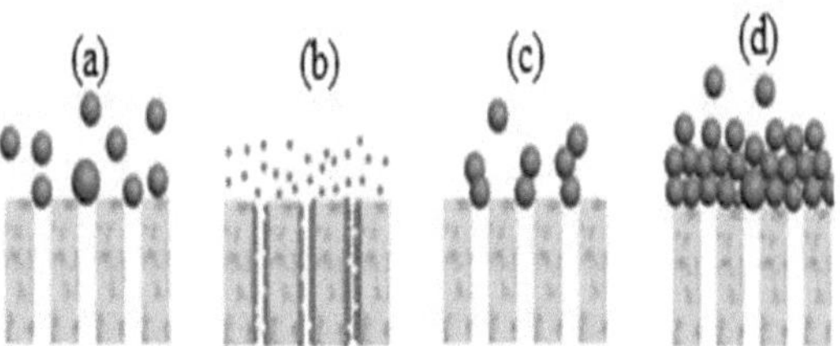

Source: Wang et al. (2020).

## 2.3. Performance of the Dynamic Membrane Anaerobic Bioreactor in Wastewater Treatment

The first research into the application of dynamic membrane anaerobic bioreactors in wastewater treatment took place in the mid-1990s (PILLAY et al., 1994). However, since 2010, research has been carried out with the aim of evaluating the bioreactor's performance in anaerobic systems for treating synthetic and domestic wastewater, emphasising the removal efficiencies of organic matter and nutrients, the production of biogas and also analysing the various factors that influence the formation of the dynamic membrane and the efficiency of the bioprocess.

Table 2 shows some studies involving the use of BRAnMD on bench and pilot scales. In general, the results show that the efficiency of the organic matter removal process is independent of the system configuration (external or submerged module), but is related to the organic load applied, the type of substrate and the operating temperature. For applied organic loads varying between 0.66 and 5.5

kgDQO.m$^{-3}$ .d$^{-1}$ , the organic matter removal efficiency in terms of total COD (DQOt) reached values between 63 and 90 per cent for domestic wastewater and between 75 and 99 per cent for synthetic wastewater. In terms of soluble COD removal efficiency, there was a variation between 39 and 78 per cent and between 65 and 98.5 per cent for domestic and synthetic wastewater, respectively. Thus, comparing these values to those presented by other anaerobic systems, such as the UASB reactor treating domestic sewage, it can be considered that the BRAnMD presented satisfactory results. However, for applied organic loads equal to or greater than 6.0 kgDQO.m$^{-3}$ .d$^{-1}$ , the removal efficiencies of total and soluble COD from domestic wastewater were considerably reduced (60 and 15.2 per cent, respectively), to the point where the system is considered to have low efficiency (YANG et al., 2020). These results can be explained by the fact that systems with high organic loads result in high concentrations of soluble microbial products (PMS) in the liquid phase, and inorganic substances and protein-like substances in the dynamic layer, resulting in a denser and more compact membrane (severe fouling). In these cases, there can be a significant reduction in the degradation and filtration of pollutants from the medium, leading to the loss of microbial metabolites and an increase in the concentration of organic matter in the effluent.

With regard to the removal of nutrients such as phosphorus and nitrogen, the low values normally presented (Table 2) may have been due to the fact that nitrogen and phosphorus are soluble in water, and small soluble molecules are weakly adsorbed by sludge particles, thus resulting i n  low adhesion (ERSHAIN et al., 2014; WANG et al., 2018). However, Ma et al. (2013b) found nitrogen and phosphorus removal efficiencies of 30 and 62 per cent, respectively. The authors used additional equipment for nutrient recovery in order to optimise the efficiency of the bioprocess.

**Table 2.** Performance of dynamic membrane anaerobic bioreactors in wastewater treatment.

| Configuration | Substrate | System volume (L) | Temperature (o C) | Organic load (kgDQO.m-3 .d)-1 | Permeate flow (L.m-2.h -1) | Organic matter removal efficiency (%) | Nutrient removal efficiency (%) | Biogas production [mL CH4.(g DQOremo life) ]-1 | References |
|---|---|---|---|---|---|---|---|---|---|
| MS | ARS | 11 | ND | MD | 137,1 | 89.5 - COD | 21.5 - NH3 | ND | Zhao et al. (2010) |
| MS | ARD | 45 | 10 a 30 | ND | 65 | 63.4 - COD | ND | ND | Zhang et al. (2011) |
| MS | ARD | 2,0 | 7,5 a 30 | ND | 60 | 79.4 - COD | 60 - Pt | ND | Ma et al. (2013a) |
| MS | ARD | 2,0 | 7,5 a 30 | ND | 60 | 81.6 - COD | 30,3 - NH3 62 - Pt | ND | Ma et al. (2013b) |
| ME | ARS | 0,898 | 35 ± 1 | 0,73 a 5,0 | 1,0 a 7,2 | 75±8 - COD | ND | ND | Alibard i et al. (2014) |
| ME | ARS | 0,898 | 20 a 24 | 0,16 a 3,3 | 1,4 a 28,0 | 80 - CODt 90 - DQOs | ND | 85 | Alibard i et al. (2016) |

| Configuração | Substrato | Volume do sistema (L) | Temperatura (°C) | Carga orgânica (kgDQO.m⁻³.d⁻¹) | Fluxo do permeado (L.m⁻².h⁻¹) | Eficiência de remoção de matéria orgânica (%) | Eficiência de remoção de nutrientes (%) | Produção de biogás [mL CH₄.(g DQOremovida)⁻¹] | Referências |
|---|---|---|---|---|---|---|---|---|---|
| MS | ARS | 6,8 | 35,7 ± 1 | 2,0 | 2,6 | 99 - CODt 63 - DQOs | 20 - Nt 13 - Pt | 310 | Ersahin et al. (2014) |
| MS | ARS | 7,4 | 35,7 ± 0,1 | 2,0 | 2,2 | 99 - COD 65 - DQOs | 18 - NT 16 - Pt | 280 a 310 | Ersahin et al. (2016a) |
| MS | ARS | 7,4 | 35,5 ± 0,2 | 2,0 | 2,2 | 99 - COD 65 - DQOs | ND | ND | Ersahin et al. (2016b) |
| MS | ARD | 3,5 | 20 | 0,88 a 3,01 | 22,5 | 70-90 - CODt 50-70 - DQOs | ND | ND | Hu et al. (2018a) |
| MS | ARS | 14,0 | 25 | ND | 50 a 150 | 80 - CODt | 10 - Nt 30 - Pt | ND | Wang et al. (2018) |

**Table 2.** Performance of dynamic membrane anaerobic bioreactors in wastewater treatment.

Continuação...

| Configuração | Substrato | Volume do sistema (L) | Temperatura (°C) | Carga orgânica (kgDQO.m⁻³.d⁻¹) | Fluxo do permeado (L.m⁻².h⁻¹) | Eficiência de remoção de matéria orgânica (%) | Eficiência de remoção de nutrientes (%) | Produção de biogás [mL CH₄.(g DQOremovida)⁻¹] | Referências |
|---|---|---|---|---|---|---|---|---|---|
| MS | ARS | 4,52 | 35 | 2,34 a 5,2 | 36 a 51,5 | 75 a 89 - DQOt | ND | 260 | Paçal et al. (2019) |
| ME | ARS | 10,0 | 37 ± 1 | 1 a 5,5 | 14 a 28 | 93 - DQOt 98,5 - DQOs | ND | ND | Berkessa et al. (2020) |
| ME | ARD | 62,8 | 19 a 30 | 1,43 | 780 | 86 a 88 - DQOt 74 a 78 - DQOs | 20,3 - Nt 16,4 - Pt | 78 a 91 | Chimuca et al. (2020) |
| ME | ARD | 62,8 | 19 a 30 | 1,43 | 780 | 86 a 88 - DQO 74 a 78 - DQOs | 20,3 - Nt 16,4 - Pt | 78 to 91 | Chimuca et al. (2021) |
| MS | ARD | 3,6 | 20 a 25 | 0,82 a 3,01 | 22,5 a 90 | 71 a 75 - DQOt 39 - DQOs | ND | 80 a 120 | Yang et al. (2020) |
| MS | ARD | 3,6 | 20 a 25 | 6,8 | 180 | 60 - DQOt 15 - DQOs | ND | 50 | Yang et al. (2020)[1] |
| MS | ARS | 6,0 | 33 ± 2 | ND | 5 a 15 | 95 - DQOt | ND | ND | Yurtsever et al. (2020) |

Where: ME: External module; MS: Submerged module; ARD: Domestic wastewater; ARS: Synthetic wastewater; CODt: Total chemical oxygen demand; CODOs: Soluble chemical oxygen demand; Pt: Total phosphorus; Nt: Total nitrogen; ND: Not available.

Analysing these results, Hu et al. (2020) associated the good performance of BRAnMD, especially in the treatment of wastewater with a low concentration of solids, with room temperature. In this condition, since the phosphorus is in particulate form, it can be easily assimilated by the sludge and/or accumulated in the stable dynamic layer. On the other hand, nitrogen removal efficiencies are relatively lower because ammoniacal nitrogen, the main component of total nitrogen in wastewater, cannot be oxidised by anaerobic microorganisms and therefore leaves the reactor dissolved in the liquid effluent. Although it is not possible to achieve high nutrient removal efficiencies in anaerobic systems, in BRAnMD removal can occur through chemical and physical mechanisms. The membrane cake layer can have high concentrations of phosphorus and nitrogen, making it feasible to recover these elements in the form of struvite (natural fertiliser), which occurs in the presence of appropriate concentrations of ions such as calcium, magnesium and potassium and under alkaline conditions. Precipitation occurs when the combined concentrations of $Mg^{2+}$, $NH_4^+$ and $PO_4^{3-}$ exceed the solubility product of the struvite, which in turn is controlled by the pH of the medium. As pH increases, solubility decreases, which makes pH one of the most important factors for precipitation of this material (MARTÍ et al., 2010; JUNG and LOVITT, 2011; CASTRO et al, 2015; ERSAHIN et al., 2017).The concentration of orthophosphate in the medium is another important factor. According to Britton et al. (2005), only at an orthophosphate concentration equal to or greater than 40 $mg.L^{-1}$ is it possible to maintain a high formation efficiency in the form of struvite. However, the product of the molar concentrations of the constituents that form struvite present in the liquid phase of the anaerobic digester is not sufficient to induce the formation of a substantial quantity of the product, limiting its efficiency in terms of phosphorus recovery. This is why, in the BRAnMD system, removal takes place as a result of the growth of the dynamic membrane and the reduction in pores, making it more efficient to recover phosphorus. that the particulate material is retained in the membrane itself. It is this fact that makes dynamic membrane bioreactors such an interesting alternative because, as well as carrying out secondary treatment efficiently, they also allow the recovery of nutrients contained in the cake, thus avoiding possible tertiary treatment.With regard to biogas production, it was observed that when the organic load ranged from 0.66 to 5.5 $kgDQO.m^{-3}$ $.d^{-1}$ in synthetic wastewater (with the exception of ALIBARDI et al, 2016), methane yields were high and ranged from 260 to 310 $mLCH_4$ $.(gDQOremoved)^{-1}$ , which was equivalent to 66 and 86 per cent of the theoretical value, respectively (395 $mLCH_4$ $.(gDQOremoved)^{-1}$ , at 35 °C). On the other hand, in domestic wastewater, methane yields were relatively lower and ranged from 80 to 120 $mLCH_4$ $.(gDQOremovida)^{-1}$ , reaching, respectively, 22.9 to 34.3 % of the theoretical value equivalent to 350 $mLCH_4$ $.(gDQOremovida)^{-1}$ , which was measured under Normal Temperature and Pressure Conditions (CNTP). And for organic loads equal to or greater than 6.0 $kgDQO.m^{-3}$ $.d^{-1}$ , biogas production was even lower (50 $mLCH_4$ $.(gDQOremoved)^{-1}$ , i.e. 14.3 per cent of the theoretical value measured at CNTP), since the efficiency of organic matter degradation was also lower. Hu et al. (2018b) state that the lower biogas yields for domestic sewage, when compared to those obtained for synthetic and industrial wastewater, can be explained by the fact that in domestic wastewater, particulate substances account for more than 50 per cent of total organic matter. In contrast, synthetic wastewater contains dissolved and easily biodegradable organic substances (such as glucose and acetate). These substances can be effectively degraded by anaerobic biomass under well-controlled operating conditions, thus contributing to high organic matter removal and, consequently, higher biogas production. In addition, domestic wastewater can contain some refractory organic compounds (even inhibitory and toxic substances) or of moderate biodegradability, which reduce COD removal rates and can inhibit biogas production, while also compromising the operating stability of the BRAnMD (HU et al., 2020; YANG et al., 2020).

Alibardi et al. (2016) found a relatively low methane yield (85 mLCH$_4$ .(gDQOremoved)$^{-1}$ ), even when treating synthetic wastewater. The authors found that, due to the configuration of the BRAnMD, part of the biogas produced was dissolved in the liquid medium and thus left the bioreactor with the effluent.

A similar behaviour was observed by Chimuca et al. (2020) treating domestic wastewater. The authors attributed the low yield to the dissolution of biogas in the liquid medium, due to the supersaturation conditions of the BRAnMD system as the DM reached irreversible fouling. The authors also found that the system's supersaturation conditions occur considerably when backwashing is not applied.

On the other hand, analysing biogas production in terms of temperature (Table 2), it was found that the high methane yields (260 e 310 mLCH$_4$ .(gDQOremoved)$^{-1}$ were obtained at higher and more constant temperatures (around 35° C). In contrast, all the low methane yields were obtained at relatively lower and variable temperatures (between 19° C and 30° C) during the operation of the respective systems (including ALIBARDI et al., 2016). This fact proves that temperature significantly influences gas-liquid mass transfer, since the solubility of biogas in liquid is inversely proportional to the operating temperature of the bioreactor. In addition, the gas-liquid mass transfer coefficient changes significantly depending on the reactor configuration and operating conditions, and can lead to methane concentrations in the liquid phase up to 12 times higher than the thermodynamic equilibrium values (PAUSS et al., 1990). This shows the diversity of possibilities offered by the application of dynamic membrane bioreactors in the anaerobic treatment of wastewater. This is because, in addition to stabilising the organic matter in the effluent and eliminating the need for a secondary decanter, it also allows for the sustainable and economically profitable use of the by-products generated.

## 2.4. Prospects for using BRAnMD for wastewater treatment: Future challenges

Studies carried out over the years show that the BRAnMD system has shown excellent results in terms of removing organic matter (expressed in terms of COD) and total suspended solids, providing low turbidity to the permeate (effluent) equivalent to that characteristic of ultra and microfiltration membrane bioreactors. On the other hand, the removal of nutrients (N and P) has been relatively low in some studies (ALIBARDI et al., 2014, 2016; ERSHAIN et al., 2014, 2016b, 2017; WANG et al., 2018; HU et al., 2018a,b), just as biogas production and the average methane yield are usually lower than their theoretical values, due to the considerable output of dissolved biogas in the system effluent and the operating conditions that are adopted (ALIBARDI et al., 2016; ERSAHIN et al., 2016b; HU et al., 2018b). Therefore, there is still a need to improve the configuration and operating conditions of the system to at least optimise biogas production and methane yield, in order to make BRAnMD more sustainable. The numerous studies carried out to date using BRAnMD have focused on operating parameters (PTM, permeate flow, critical flow, TDH and TRC), the structure of the dynamic layer (physico-chemical and biological properties of the cake layer), the morphological properties of the sludge (sedimentation, dehydration, flocculability, hydrophobicity, SPE and PMS), the quality of the permeate produced (COD, turbidity, solids, N and P) and modelling the formation of the dynamic membrane and the mechanisms responsible for fouling, such as the XDLVO Theory (MA et al., 2013a,b; ALIBARDI et al., 2014, 2016; CHU et al., 2014; ERSAHIN et al.., 2014, 2016a,b, 2017; LIU et al., 2016; HU et al., 2016, 2017, 2018a,b; GUAN et al., 2018a,b; WANG et al., 2018; SIDDIQUI et al., 2019; YU et al.; 2019a,b).However, there are still challenges to be overcome and answers to be found, such as obtaining a more complete mathematical model that can add (concurrently or not) operational a n d  physicochemical parameters, thermodynamic and hydrodynamic forces and biological properties of the sludge in order to accurately determine the critical flow and backwashing frequencies. In this way, the operation of the BRAnMD would have greater control in the medium and long term, and problems such as fouling could be avoided. Furthermore, very few studies have evaluated biological parameters such as helminth eggs (CHIMUCA et al., 2020).Most of the research carried out has been on a bench scale and with temperature control. It is therefore still a major challenge to develop, build and monitor full-scale plants operated at room temperature, with the aim of applying the dynamic membrane anaerobic bioreactor in individual and collective domestic and industrial wastewater treatment systems. In addition, much still needs to be clarified in terms of resource removal and reuse, which is why basic research and full-scale projects need to be carried out continuously in order to achieve increasingly optimised systems.

# 3. CONCLUSION

Dynamic membrane anaerobic bioreactors have provided results that are on a par with those obtained in conventional MF and UF membrane systems, presenting effluents with low turbidity and high removal efficiencies for organic matter, suspended solids and helminth eggs. In addition, the dynamic membrane is formed on a low-cost support material and filtration can take place by gravity, which further reduces treatment costs.With regard to fouling, although it causes an increase in PTM and a reduction i n  permeate flow, it is a problem that can be easily solved by, for example, backwashing using the treated effluent itself. The fouling-free support material allows the dynamic membrane to be formed again and the effectiveness of the process to be recovered.In addition, the formation of the dynamic membrane, which is crucial for the good performance of the process, can be explained by various theories, which can be based on hydrodynamic and thermodynamic forces, as well as the speeds at which the sludge flakes reach the support material. The process, which normally takes place in two stages (initial and maturation), must allow for the formation of a robust, dense and stable membrane with suitable permeability and filterability characteristics. However, despite the numerous positive points, a lot still needs to be investigated so that the obstacles and doubts that exist regarding the use of BRAnMD in wastewater treatment can be eliminated. The necessary advances must seek measures that further reduce process costs, making this type of bioreactor increasingly competitive.

# 4. REFERENCES

ALIBARDI, L., BERNAVA, N., COSSU, R, SPAGNI, A. (2016)
Anaerobic dynamic membrane bioreactor for wastewater treatment at
ambient temperature. Chemical Engineering Journal, **284**, 130-138.
doi: 10.1016/j.cej.2015.08.111

ALIBARDI, L., COSSU, R., SALEEM, M., SPAGNI, A. (2014)
Development
and permeability of a dynamic membrane for anaerobic wastewater
treatment. Bioresource Technology, **161**, 236-244. doi:
10.1016/j.biortech.2014.03.045

AN, Y., WANG, Z., WU, Z., YANG, D., ZHOU, Q. (2009)
Characterisation of membrane foulants in an anaerobic non-woven
fabric membrane bioreactor for municipal wastewater treatment.
Chemical Engineering Journal, **155**, 709-715. doi:
10.1016/j.cej.2009.09.003

AYOL, A., DEMIRAL, Y.O., GÜNEŞ, S. (2021) Efficient treatment
of domestic wastewaters by using a dynamic membrane bioreactor
System. Journal of Membrane Science & Research, **7**, 55-58. doi:
10.22079/JMSR.2020.120244.1330

AZEREDO, J., VISSER, J., OLIVEIRA, R. (1999) Exopolymers in
bacterial adhesion: interpretation in terms of DLVO and XDLVO
theories. Colloids and Surfaces B : Biointerfaces, **14**, 141-148, doi:
10.1016/S0927-7765(99)00031-4

BERKESSA, Y.W., YAN, B., LI, T., JEGATHEESAN, V., ZHANG,
Y. (2020)
Treatment of anthraquinone dye textile wastewater using anaerobic

dynamic membrane bioreactor: performance and microbial dynamics. Chemosphere, **238**, 1-11. doi: 10.1016/j.chemosphere.2019.124539

BÉRUBÉ, P., HALL, E.R., SUTTON, P.M. (2006) Parameters governing permeate flux in an anaerobic membrane bioreactor treating low-strength municipal wastewaters: A literature review. Water Environment Research, **78,** 887-896. doi: 10.2175/106143005X72858

Britton et al. (2005)

CAI, D., HUANG, J., LIU, G., LI, M., YU, Y., MENG, F. (2018) Effect of

support material pore size on the filtration behaviour of dynamic membrane bioreactor. Bioresource Technology, **255**, 359-363. doi: 10.1016/j.biortech.2018.02.007

CASTRO, S.R., CRUTCHIK, D., GARRIDO, J.M., LANGE, **L** . C. (2015)

Chemical precipitation of struvite: nutrient recovery in a conical fluidised bed reactor using low-cost industrial magnesia. Sanitary and Environmental Engineering, **20**(21), 259-268. doi: 10.15 90/S1413-41522015020000133827

CHANG, I-S., JUDD, S.J. (2002) Air sparging of submerged MBR for municipal wastewater treatment. Process Biochemistry, **37,** 915-920. doi: 10.1016/S0032-9592(01)00291-6

CHIMUCA, J.F.J., SOUSA, J.T, LOPES, W.S., LEITE, V.D., CANTO, C.S.A. (2020) Decentralised treatment of domestic sewage in dynamic membrane bioreactor. Desalination and Water Treatment, **197**, 76-89 doi: 10.5004/dwt.2020.25981

CHIMUCA, J.F.J., SOUSA, J.T, LOPES, W.S., CANTO, C.S.A, LEITE,V.D (2021) Fouling mechanism in dynamic membrane anaerobic bioreactor treating domestic sewage: filtration performance. Desalination and Water Treatment, **236**, 26-44. doi: 10.5004/dwt.2021.27684

CHU, H-Q., CAO, D-W., JIN, W., DONG, B-Z. (2008) Characteristics of bio-diatomite dynamic membrane process for municipal wastewater treatment. Journal of Membrane Science, **325**(1), 271-276. doi: 10.1016/j.memsci.2008.07.040

CHU, H., ZHANG, Y., ZHOU, X., ZHAO, Y., DONG, B., ZHANG, H. (2014) Dynamic membrane bioreactor for wastewater treatment: Operation, critical flux, and dynamic membrane structure. Journal of Membrane Science, **450,** 265-271. doi: 10.1016/j.memsci.2013.08.045

CUI, Z.F., CHANG, S., FANE, A.G. (2003) The use of gas bubbling to enhance membrane processes. Journal of Membrane Science, **221**(1-2), 1-35. doi: 10.1016/S0376-7388(03)00246-1

Davis, M.L. (2010) Water and Wastewater Engineering - Design Principles and Practice. McGraw-Hill, NY, 1301 pp.

DERELI, R.K., ERSAHIN, M.E., OZGUN, H., OZTURK, I., JEISON, D., VAN DER ZEE, F., VAN LIER, J.B. (2012) Potentials of anaerobic membrane bioreactors to overcome treatment limitations induced by industrial wastewaters. Bioresource Technology, **122**, 160-170. doi: 10.1016/j.biortech.2012.05.139

Ersahin et al. (2016a*table Ersahin et al. (2016b*table

ERSAHIN, M.E., GIMENEZ, J.B., OZGUN, H., TAO, Y., SPANJERS, H.,VAN LIER, J.B. (2016a) Gas-lift anaerobic dynamic membrane bioreactors for high strength synthetic wastewater treatment: Effect of biogas sparging velocity and HRT on treatment performance. Chemical Engineering Journal, **305,** 46-53. doi:

10.1016/j.cej.2016.02.003

ERSAHIN, M.E., GIMENEZ, J., OZGUN, H., TAO, Y., SPANJERS, H., VAN LIER, J.B. (2016b) Gas-lift anaerobic dynamic membrane bioreactors for high strength synthetic wastewater treatment: Effect of biogas sparging velocity and HRT on treatment performance. Chemical Engineering Journal, **305**, 46-53. doi: 10.1016/j.cej.2016.02.003

ERSAHIN, M.E., OZGUN, H., TAO, Y., VAN LIER, J.B. (2014) Applicability of dynamic membrane technology in anaerobic membrane bioreactors. Water Research, **48**, 420-429. doi: 10.1016/j.watres.2013.09.054 ERSAHIN, M.E., TAO, Y., OZGUN, H., GIMENEZ, J.B., SPANJERS, H.,

VAN LIER, J.B. (2017) Impact of anaerobic dynamic membrane bioreactor configuration on treatment and filterability performance. Journal of Membrane Science, **526**, 387-394. doi: 10.1016/j.memsci.2016.12.057

ERSAHIN, M.E., TAO, Y., OZGUN, H., SPANJERS, H., VAN LIER, J.B. (2016a) Characteristics and role of dynamic membrane layer in anaerobic membrane bioreactors. Biotechnology and Bioengineering, **113**(102), 761-771. doi: 10.1002/bit.25841

FAN, B., HUANG, X. (2002) Characteristics of a self-forming dynamic membrane coupled with a bioreactor for municipal wastewater treatment. Environnental Science and Technology, **36**(23), 5245-5251. doi: 10.1021/es025789n

FANE, A.G., FELL, C.J.D. (1987) A review of fouling and fouling control in ultrafiltration. Desalination, **62**, 117-136. doi: 10.1016/0011-9164(87)87013-3

FANE, A.G., WANG, R., JIA, Y. (2011) Membrane technology: past, present and future. In: Wang L.K., Chen J.P., Hung YT., Shammas N.K. (Eds), Membrane and desalination technologies. Handbook of environmental engineering, **13,** Humana Press, Totowa, NJ, 1-45. https://doi.org/10.1007/978-1-59745-278-6_1

GUAN, D., DAI, J., WATANABE, Y., CHEN, G. (2018b) Changes in the physical properties of the dynamic layer and its correlation with permeate quality in a self-forming dynamic membrane bioreactor. Water Research, **140**, p. 67-76. doi: 10.1016/j.watres.2018.04.041

GUAN, D., SIDDIQUI, M., CHEN, G. (2018a) Comparison of different chemical cleaning reagents on fouling recovery in a self-forming dynamic membrane bioreactor (SFDMBR). Separation and Purification Technology, **206**, 158-165. doi: 10.1016/j.seppur.2018.05.059

HAN, S-S., BAE, T-H., JANG, G-G., TAK, T-M. (2005) Influence of sludge
retention time on membrane fouling and bioactivities in membrane bioreactor system. Process Biochemistry, **40**(7), 2393-2400. doi: 10.1016/j.procbio.2004.09.017

HE, Y., XU, P., LI, C., ZHANG, B. (2005) High-concentration food wastewater treatment by an anaerobic membrane bioreactor. Water Research, **39**(17), 4110-4118. doi: 10.1016/j.watres.2005.07.030

HONG, H., PENG, W., ZHANG, M., CHEN, J., HE, Y., WANG, F. (2013)
Thermodynamic analysis of membrane fouling in a submerged membrane bioreactor and its implications. Bioresource Technology, **146**, 7-14. doi: 10.1016/j.biortech.2013.07.040

HU, Y., WANG, X.C., NGO, H.H., SUN, Q., YANG, Y. (2018b) Anaerobic dynamic membrane bioreactor (AnDMBR) for wastewater treatment: A review. Bioresource Technology, **247**, 1107-1118. doi: 10.1016/j.biortech.2017.09.101

HU, Y., WANG, X.C., SUN, Q., NGO, H.H., YU, Z., TANG, J. (2017)

Characterisation of a hybrid powdered activated carbon-dynamic membrane bioreactor (PAC-DMBR) process with high flux by gravity flow: operational performance and sludge properties. Bioresource Technology, **223,** 65-73. doi: 10.1016/j.biortech.2016.10.036

Hu et al. (2016a)

HU, Y., WANG, X.C., TIAN, W., NGO, H.H., CHEN, R. (2016a) Towards stable operation of a dynamic membrane bioreactor (DMBR): operational process, behaviour and retention effect of dynamic membrane. Journal of Membrane Science, **498**, 20-29. doi: 10.1016/j.memsci.2015.10.009

HU, Y., YANG, Y., YU, S., WANG, X.C., TANG, J. (2018a) Psychrophilic

anaerobic dynamic membrane bioreactor for domestic wastewater treatment: Effects of organic loading and sludge recycling. Bioresource Technology, **270,** 62-69. doi: 10.1016/j.biortech.2018.08.128

HU, Y., YANG, Y., ZANG, Y., ZHANG, J., WANG, X.C. (2020) 11- Anaerobic dynamic membrane bioreactors (AnDMBRs) for wastewater treatment. Current Development in Biotechnology and Bioengineering, 259-281. doi: 10.1016/B978-0-12-819852-0.00011-7

HUANG, J., WU, X., CAI, D., CHEN, G., LI, D., YU, Y., PETRIK, L.F., LIU.

G. (2019) Linking solids retention time to the composition, structure, and hydraulic resistance of biofilms developed on support materials in dynamic membrane bioreactors. Journal of Membrane Science, **581**, 158-167. doi: 10.1016/j.memsci.2019.03.033

HUANG, Z., ONG, S.L., NG, H.Y. (2011) Submerged anaerobic membrane bioreactor for low-strength wastewater treatment: Effect of HRT and SRT on treatment performance and membrane fouling. Water Research, **45**(2), 705-713. doi: 10.1016/j.watres.2010.08.035

IRITANI, E., KATAGIRI. N. (2016) Developments of blocking filtration model in membrane filtration. Kona Powder and Particle Journal, **33**, 179-202. doi: 10.14356/kona.2016024

KANG, I-J., YOON, S-H., LEE, C-H. (2002) Comparison of the filtration characteristics of organic and inorganic membranes in a membrane-coupled anaerobic bioreactor. Water Research, **36**(7), 1803-1813. doi: 10.1016/S0043-1354(01)00388-8

KISO, Y., JUNG, Y.J., MIN, K.S., WANG, W., SIMASE, M., YAMADA, T., MIN, K.S., (2005) Coupling of sequencing batch reactor and mesh filtration: operational parameters and wastewater treatment performance. Water Research, **39**(20), 4887-4898. doi: 10.1016/j.watres.2005.05.025

LADEWIG, B., AL-SHAELI, M.N.Z. (2017) Fundamentals of Membrane Bioreactors - Materials, Systems and Membrane Fouling, Springer, Singapore, 155 pp.

LEE, W., KANG, S, SHIN, H. (2003) Sludge Characteristics and their contribution to microfiltration in submerged membrane bioreactors. Journal of Membrane Science, **216**(1-2), 217-227. doi: 10.1016/S0376-7388(03)00073-5

LI, L., XU, G., YU, H. (2018) Dynamic membrane filtration: Formation, filtration, cleaning, and applications. Chemical

Engineering and Technology, **41**(1), 7-18. doi: 10.1002/CEAT.201700095

LI, L., XU, G., YU, H., XING, J. (2018b) Dynamic membrane for micro-particle removal in wastewater treatment: Performance and influencing factors. Science of the Total Environment, **627**, 332-340. doi: 10.1016/j.scitotenv.2018.01.239

LIANG, S., LIU, C., SONG, L. (2007) Soluble microbial products in membrane bioreactor operation: Behaviors, characteristics, and fouling potential. Water Research, **41**(1), 95-101. doi: 10.1016/j.watres.2006.10.008

LIAO, B.Q., BAGLEY, D.M., KRAEMER, H.E., LEPPARD, G.G., LISS, S.N.A. (2004) Review of biofouling and its control in membrane separation bioreactors. Water Environment Research, 76(5), 425-436. doi: 0.2175/106143004X151527

LIN, H., PENG, W., ZHANG, M., CHEN, J., HONG, H., ZHANG, Y. (2013) A review on anaerobic membrane bioreactors: Applications, membrane fouling and future perspectives. Desalination, **314**, 169-188. doi: 10.1016/j.desal.2013.01.019

LIU,Y.,LIU,H.,CUI,L.,ZHANG, K.(2012) The ration of food-to-microorganism (F/M) on membrane fouling of anaerobic membrane bioreactors treating low-strength wastewater. Desalination, **297**, 97-103. doi: 10.1016/j.desal.2012.04.026

LIU, H., WANG, Y., YIN, B., ZHU, Y., FU. B., LIU, H. (2016) Improving volatile fatty acid yield from sludge anaerobic fermentation through self-forming dynamic membrane separation. Bioresource Technology, **218,** 92-100. doi: 10.1016/j.biortech.2016.06.077

LIU, H., YANG, C., PU, W., ZHANG, J. (2009) Formation
mechanism and structure of dynamic membrane in the dynamic
membrane bioreactor. Chemical Engineering Journal, **148**, 290-295.
doi: 10.1016/j.cej.2008.08.043

MA,J.,WANG,Z.,ZOU,X.,FENG,J.,WU,Z.(2013a) Microbial
communities in an anaerobic dynamic membrane bioreactor
(AnDMBR) for municipal wastewater treatment: Comparison of bulk
sludge and cake layer. Process Biochemistry, **48**, 510-516. doi:
10.1016/j.procbio.2013.02.003

MA. J., WANG, Z., XU, Y., WANG, Q., WU, Z., GRASMICK, A.
(2013b)
Organic matter recovery from municipal wastewater by using dynamic
membrane separation process. Chemical Engineering Journal, **219**,
190-199. doi: 10.1016/j.cej.2012.12.085

MAHAT, S.B., OMAR, R., IDRIS, A., KAMAL, S.M.M., IDRIS,
A.I.M. (2018)
Dynamic membrane applications in anaerobic and aerobic digestion
for industrial wastewater: A mini review. Food and Bioproducts
Processing, **12**, 150-168. doi: 10.1016/j.fbp.2018.09.008

MARCINKOWSKY, A.E., KRAUS, K.A., PHILLIPS, H.O.,
JOHNSON JR., J.S., SHOR, A.J. (1966) Hyperfiltration studies. IV.
Salt rejection by dynamically formed hydrous oxide membranes.
Journal of the American Chemical Society, **88**(24), 5744-5746.

MARTÍ, N., PASTOR, L., BOUZAS, A., FERRER, J., SECO, A.
(2010)
Phosphorus recovery by struvite crystallisation in WWTP's: Influence
of the sludge treatment line operation. Water Research, **44**(7), 2371-
2379. doi: 10.1016/j.watres.2009.12.043

MARTINEZ-SOSA, D., HELMREICH, B., HORN, H. (2012) Anaerobic

submerged membrane bioreactor (AnSMBR) treating low-strength wastewater under psychrophilic temperature conditions. Process Biochemistry, **47**(5), 792-798. doi: 10.1016/j.procbio.2012.02.011

MEABE, E., DÉLÉRIS, S., SOROA, S., SANCHO, L. (2013) Performance of

anaerobic membrane bioreactor for sewage sludge treatment: Mesophilic and thermophilic processes. Journal of Membrane Technology, **446**, 26-33. doi: 10.1016/j.memsci.2013.06.018

MENG, F., ZHANG, Y., OH, Z., ZHOU, H., SHIN, S-H, CHAE, S-R. (2017)

Fouling in membrane bioreactors: An updated review. Water Research,

**114**, 151-180. doi: 10.1016/j.watres.2017.02.006

MENG, F., CHAE, S.R., DREWS, A., KRAUME, M., SHIN, H.S., YANG, F.

(2009) Recent advances in membrane bioreactors (MBRs): membrane fouling and membrane material. Water Research, **43**(6), 1489-1512. doi: 10.1016/j.watres.2008.12.044

METCALF, L., EDDY, H.P., TCHOBANOGLOUS, G. (2007) Wastewater

engineering: treatment, disposal, and reuse. New York: McGraw-Hill, 1. ed., 1503 pp.

NG, H. Y., HERMANOWICZ, S. W. (2005) Membrane bioreactor operation at short solids retention times: Performance and biomass characteristics. Water Research, **39**(6), 981-992. doi: 10.1016/j.watres.2004.12.014

NOOR, M.J.M.M., AHMADUN, F.R., MOHAMED, T.A., MUYIBI, S.A., PESCOD, M.B. (2002) Performance of flexible membrane using kaolin dynamic membrane in treating domestic wastewater. Desalination, **147**, 263-268. doi: 10.1016/S0011-9164(02)00548-9

OGNIER, S., WISNIEWSKI, C., GRASMICK, A. (2004) Membrane bioreactor
fouling in sub-critical filtration conditions: A local critical flux concept. Jounal Membrane of Science, **229**(1-2), 171-177. doi: 10.1016/j.memsci.2003.10.026

PAÇAL, M., SEMERCI, N., ÇALLI, B. (2019) Treatment of synthetic wastewater and cheese whey by the anaerobic dynamic membrane bioreactor. Environmental Science and Pollution Research, **26**, 32942-32956. doi: 10.1007/s11356-019-06397-z

PAJOOH, Y. Dead-end membrane module. 2018. Available at: <http://www.yasinpajooh.com/page/en-80/Dead-End-Membrane-Module>. Accessed on 25 February 2021.

PARK, H-D., CHANG, I-S., LEE, K-J. (2015) Principles of Membrane Bioreactors for Wastewater Treatment, CRC Press, NY, 436 pp.

PAUSS, A., ANDRE, G., PERRIER, M., GUIOT, S. R. (1990) Liquid-to-gas mass transfer in anaerobic processes: Inevitable transfer limitations of methane and hydrogen in the biomethanation process. Applied and Environmental Microbioloby, **56**(6), 1636-1644. doi: 10.1128/aem.56.6.1636-1644.1990

PILLAY, V.L., TOWNSEND, B., BUCKLEY, C.A. (1994) Improving the performance of anaerobic digesters at wastewater treatment works: the coupled cross-flow microfiltration/digester process. Water Science and Technology, **30**(12) 329-337. doi:

10.2166/wst.1994.0632

POLLICE, A., VERGINE, P. (2020) 10 - Self-forming dynamic membrane bioreactors (SFD MBR) for wastewater treatment: Principles and applications. In: MANNINA, G., PANDEY, A., LARROCHE, C., NG,

H.Y., NGO, H.H. (Eds.),Current Developments in Biotechnology and Bioengineering. FL. 235-258. doi: 10.1016/B978-0-12-819854-4.00010-1

PRETEL, R., ROBLES, A., RUANO, M. V., SECO, A., FERRER, J. (2013)

Environmental impact of submerged anaerobic MBR (SAnMBR) technology used to treat urban wastewater at different temperatures. Bioresource Technology, **149**, 532-540.doi: 10.1016/j.biortech.2013.09.060

QUEK, P.J., YEAP, T.S, NG, H.Y. (2017) Applicability of upflow anaerobic sludge blanket and dynamic membrane-coupled process for the treatment o f  municipal wastewater. Applied Microbiology and Biotechnology, **101**, 6531-6540. doi: 10.1007/s00253-017-8358-6

RUAN, B., WU, P., LIU, J., JIANG, L., WANG, H., QIAO, J , ZHU, N., DANGA, Z., LUOA, H., YI, X. (2020) Adhesion of Sphingomonas sp. GY2B onto montmorillonite: A combination study by thermodynamics and the extended DLVO theory. Colloids and Surfaces B: Biointerfaces, **192**, 111085. doi: 10.1016/j.colsurfb.2020.111085

SCHNEIDER, R.T., TSUTIYA, M.T. Membranas Filtrantes para o Tratamento de Água, Esgoto e Água de Reúso. 1. ed., ABES, São Paulo, 2001.

SIDDIQUI, M., DAI, J.,GUAN, D., CHEN, G. (2019) Exploration of the formation of self-forming dynamic membrane in an up flow anaerobic sludge blanket reactor. Separation and Purification Technology, **212,** 757-766. doi: 10.1016/j.seppur.2018.11.065

SMITH, A. L., STADLER, L. B., LOVE, N. G, SKERLOS, S. J., RASKIN, L.

(2012) Perspectives on anaerobic membrane bioreactor treatment of domestic wastewater: a critical review. Bioresource Technology, **122,** 149-159. doi: 10.1016/j.biortech.2012.04.055

SUN, F., ZHANG, N., LIB, F., WANG, X., ZHANG, J., SONG, L., LIANG. S. (2018) Dynamic analysis of self-forming dynamic membrane (SFDM) filtration in submerged anaerobic bioreactor: Performance, characteristic, and mechanism. Bioresource Technology, **270,** 383-390. doi: 10.1016/j.biortech.2018.09.003

TOWNSEND, R.B., CRAWDRON, M.P.R., NEYTZELL, F.G., BUCKLEY, C.A. (1989) The potential of dynamic membranes for the treatment of industrial effluents. ChemSA, **15**(4), 132-134.

TRUSSELL, R.S., MERLO, R.P., HERMANOWICZ, W., JENKINS, D. (2006) The Effect of organic loading on process performance and membrane fouling in a submerged membrane bioreactor treating municipal wastewater. Water Research, **40**(14), 2675-2683. doi: 10.1016/j.watres.2006.04.020

VERGINE, P., SALERNO, C., BERARDI, G., POLLICE.A.(2021) Self-Forming dynamic membrane bioreactors (SFD MBR) for municipal wastewater treatment: relevance of solids retention time and biological process stability. Separation and Purification Technology, **255**, 1-8. doi: 10.1016/j.seppur.2020.117735

VILINSKA, A., RAO, K.H. (2011) Surface thermodynamics and

extended DLVO theory of Leptospirillum ferrooxidans cells' adhesion on sulfide minerals. Minerals & Metallurgical Processing, **28**(3), 151-158. doi: 10.1007/BF03402248

WANG, J., CAHYADI, A., WU, B., PEE, W., FANE, A.G., CHEW. J.W.

(2020) The roles of particles in enhancing membrane filtration: A review. Journal of Membrane Science, **595**, 1-29. doi: 10.1016/j.memsci.2019.117570

WANG, L., LIU, H., ZHANG, W., YU, T., JIN Q., FU, B., LIU, H. (2018)

Recovery of organic matters in wastewater by self-forming dynamic membrane bioreactor: Performance and membrane fouling. Chemosphere, **203**, 123-131. doi: 10.1016/j.chemosphere.2018.03.171

WEI, X., WANG, R., FANE, A.G. (2006) Development of a novel electrophoresis-UV grafting method to modify PES UF membranes used for NOM removal. Journal of Membrane Science, **273**, 47-57. doi: 10.1016/j.memsci.2005.11.049

WU, Y., HUANG, X., WEN, X., CHEN, F. (2005). Function of dynamic membrane in self-forming dynamic membrane coupled bioreactor. Water Science and Technology, **51**(6-7), 107-114. doi:10.2166/wst.2005.0628

YAMAGIWA, K., OOHIRA, Y., OHKAWA, A. (1994) Performance evaluation of a plunging liquid jet bioreactor with cross flow filtration for small scale treatment of domestic wastewater. Bioresource Technology, **50,** 131-138. doi: 10.1016/0960-8524(94)90065-5

YANG, Y., ZANG, Y., HU, Y., WANG, X. C., NGO. H. H. (2020) Upflow anaerobic dynamic membrane bioreactor (AnDMBR) for wastewater treatment at room temperature and short HRTs: Process characteristics and practical applicability. Chemical Engineering Journal, **383**, 1-10. doi: 10.1016/j.cej.2019.123186

YU, K., WEN, X., BU, Q., XIA, H. (2003) Critical flux enhancements with air sparging in axial hollow fibres cross-flow microfiltration of biologically treated wastewater. Journal of Membrane Science, **224**(1-2), 69-79. doi: 10.1016/j.memsci.2003.07.001

YU, H-Y., XIE, Y-J., HU, M-X., WANG, J-L., WANG, S-Y., XU, Z-K. (2005) Surface modification of polypropylene microporous membrane to improve its antifouling property in MBR: $CO_2$ plasma treatment. Journal of Membrane Science, **254**(1-2), 219-227. doi: 10.1016/j.memsci.2005.01.010

YU, Z., HU, Y., DZAKPASU, M., WANG, X.C. (2019b) Thermodynamic prediction and experimental investigation of short-term dynamic membrane formation in dynamic membrane bioreactors: Effects of sludge properties. Journal of Environmental Sciences, **77**, 85-96. doi: 10.1016/j.jes.2018.06.017

YU, Z., HU, Y., DZAKPASU, M., WANG, X.C., NGO, H.H. (2019a) Dynamic membrane bioreactor performance enhancement by powdered activated carbon addition: Evaluation of sludge morphological, aggregative and microbial properties. Journal of Environmental Sciences, **75**, 73-83. doi: 10.1016/j.jes.2018.03.003

YURTSEVER, A., BASARAN, E., UCAR, D. (2020) Process optimisation and filtration performance of an anaerobic dynamic membrane bioreactor treating textile wastewaters. Journal of Environmental Management, **273**, 1-8. doi: 10.1016/j.jenvman.2020.111114

ZHANG, X., WANG, Z., WU, Z., LU, F., TONG, J., ZANG, L.

(2010)
Formation of dynamic membrane in an anaerobic membrane bioreactor for municipal wastewater treatment. Chemical Engineering Journal, **165**(1), 175-183. doi: 10.1016/j.cej.2010.09.013

ZHANG, S., QU, Y., LIU, Y., YANG, F., ZHANG, X., FURUKAWA, K., YAMADA, Y. (2005) Experimental study of domestic sewage treatment with a metal membrane bioreactor. Desalination, **177**(1-3), 83-93. doi: 10.1016/j.desal.2004.10.034

ZHANG, X., WANG, Z., WU, Z., WEI, T., LU, F., TONG, J., MAI, S. (2011) Membrane fouling in an anaerobic dynamic membrane bioreactor (AnDMBR) for municipal wastewater treatment: characteristics of membrane foulants and bulk sludge. Process Biochemistry, **2**(4), 1538-1546. doi: 10.1016/j.procbio.2011.04.002

ZHAO,L-J., ZHAO, T-T., WANG, S-G. (2010) Study on anaerobic self-forming dynamic membrane bioreactor for domestic wastewater treatment. Journal of Shandong University (Natural Science), **45**(3), 10-14.